all about mice

by howard hirschhorn

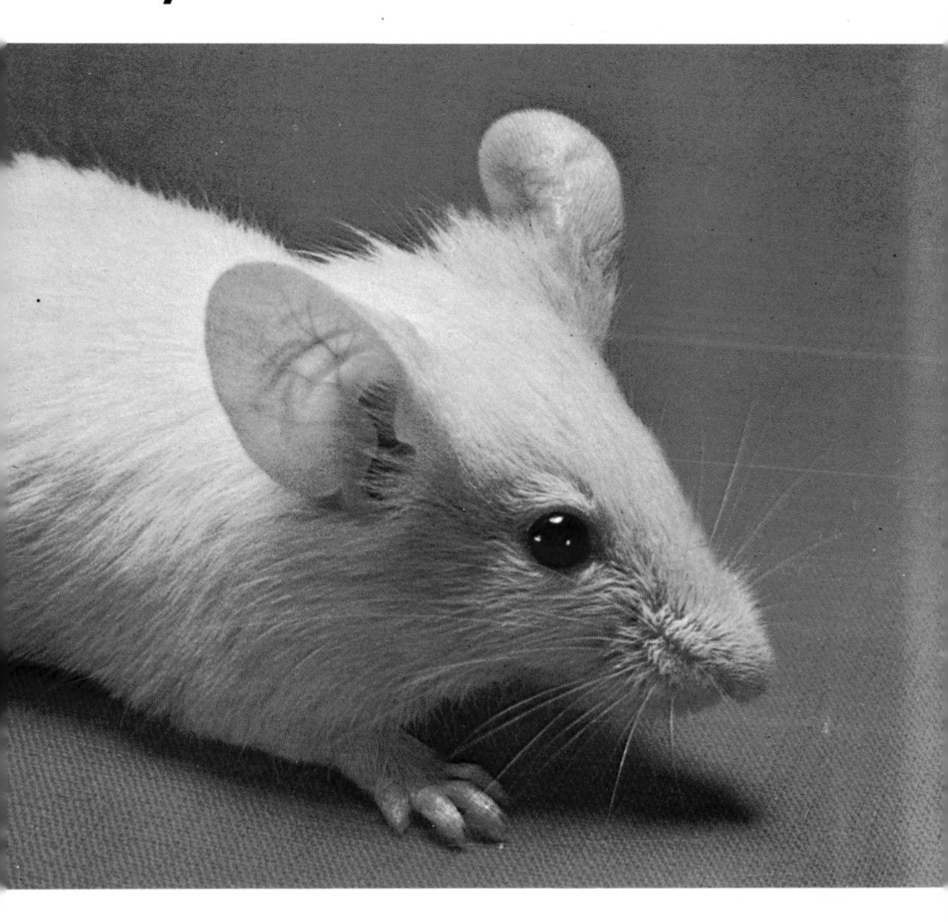

Distributed in the U.S.A. by T.F.H. Publications, Inc., 211 West Sylvania Avenue, P.O. Box 27, Neptune City, N.J. 07753; in England by T.F.H. (Gt. Britain) Ltd., 13 Nutley Lane, Reigate, Surrey; in Canada to the book store and library trade by Clarke, Irwin & Company, Clarwin House, 791 St. Clair Avenue West, Toronto 10, Ontario; in Canada to the pet trade by Rolf C. Hagen Ltd., 3225 Sartelon Street, Montreal 382, Quebec; in Southeast Asia by Y.W. Ong, 9 Lorong 36 Geylang, Singapore 14; in Australia and the south Pacific by Pet Imports Pty. Ltd., P.O. Box 149, Brookvale 2100, N.S.W., Australia. Published by T.F.H. Publications Inc. Ltd., The British Crown Colony of Hong Kong.

All photos by Marvin Apfelbaum unless otherwise credited.

Cover photograph and frontispiece, an albino mouse, the kind commonly used as pets and laboratory animals. Photo by Dr. Herbert R. Axelrod.

ISBN #0-87666-210-6

Table of Contents

"By nature timid, by necessity familiar, its fear and wants are the sole springs of its actions. It never leaves its hiding place but to seek for food; nor does it, like the rat, go from one house to another, unless forced to it; nor does it by any means cause so much mischief. The mouse is a beautiful creature; its skin sleek and soft, its eyes bright and lively; all its limbs are formed with exquisite delicacy."—From an 18th century natural history book by Comte de Buffon.

The mouse is a beautiful creature; timid by nature, but willing to be your friend if you will be its friend. Photo by Claudia Watkins, Ferguson, Mo.

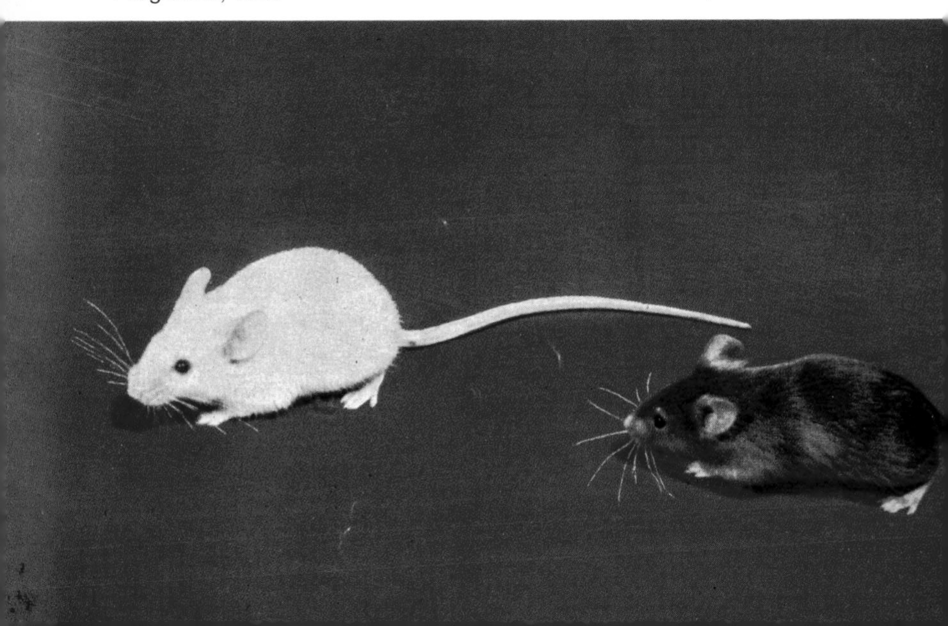

INTRODUCTION

What advantages are there in having mice as house pets? They are inexpensive to buy and maintain and do not require any profound scientific knowledge on the part of the owner. They adapt well to various conditions. They can do with very small cages and will not eat you out of house and home. They can become quite tame. Mice breed easily and are prolific (although you might not always consider this trait an advantage). And you have the added bonus of possessing livestock to compete in mouse shows! Keep this in mind when initially buying your mice; pay for just a bit more than the most common variety and you will increase your chances of selling your future litters. Better stock is just as easy and inexpensive to keep as the less expensive animals.

Indeed, a good mouse is appreciated. Even the great music chambers of nineteenth-century Europe honored the mouse. Here is a translation of an original German description of the house mouse, written by the world-renowned naturalist Alfred E. Brehm over one hundred years ago. It succinctly and charmingly depicts the real nature of this creature for us. This passage leaves us no doubt as to the musical nature of the mouse.

> *"The house mouse is a charming, extremely agile and active animal. It quickly scampers over the ground, climbs splendidly, leaps rather far and often hops around in short bounds. You can easily observe how tame mice maneuver about quite skilfully. Let a mouse run up on a diagonally stretched string or rod, then you can see it quickly wrap its tail, the moment it loses balance, around the cord or rod just like an animal with a real prehensile tail would do; as soon as the mouse regains its balance it keeps right on climbing. If you set it on a grass blade, it climbs up*

to the tip, and as soon as the blade bends under the mouse's weight, the mouse hangs on underneath and climbs slowly down without encountering any difficulty. The mouse's tail has quite a role to play when its owner climbs, for those mice whose tails have been cut off (once the style for giving a clownish look to trained mice), can no longer keep up with their companions.

"The mouse can strike all manner of poses. Every movement is a pleasure to watch. Even when it sits still—as still as a mouse—it is a happy sight. It becomes a downright bewitching creature when it raises up, as rodents do, to clean and wash itself.

"It is capable also of other clever tricks, like standing up completely on its hind legs as people do. It can even take a few steps, supporting itself a bit now and then with its tail.

"The mouse knows how to swim, too, although it only approaches water in the direst emergency. If someone happens to throw one into a pond or stream, it cuts through the water almost with the speed of a harvest mouse or a water rat. It tries to land at the nearest dry spot.

"The house mouse's senses are excellent. It hears the slightest noise, smells very keenly and for quite some distance. Its vision is also quite adequate being perhaps better at night than during the day.

"The house mouse's psyche makes it a real favorite of whomever endeavors to study its life and habits. It is good-natured and harmless, not seeming the least like the cunning and sometimes ferocious relative, the rat. It is very curious and investigates everything with the greatest care.

"It is a joyful and clever creature, and soon knows where it is allowed to live in peace, yet growing accustomed to the presence of man, even going back

and forth on its business under man's very eyes as if he were no disturbance for it at all.

"After several days in a cage, a new mouse behaves rather amicably. Even older mice finally become quite tame. The good-naturedness and harmlessness of recently captured young mice exceeds that of most other rodents which man can capture and keep as pets.

The mouse is a good-natured, joyful, clever creature which responds to living in peace with man and beast.

Mice, once they are tamed, make remarkable pets for children. They seem to be especially fond of snuggling up in dark, warm pockets of soft material.

This mouse is taken to school every day and the boy who owns him says it attracts more girls than it frightens! ⟶

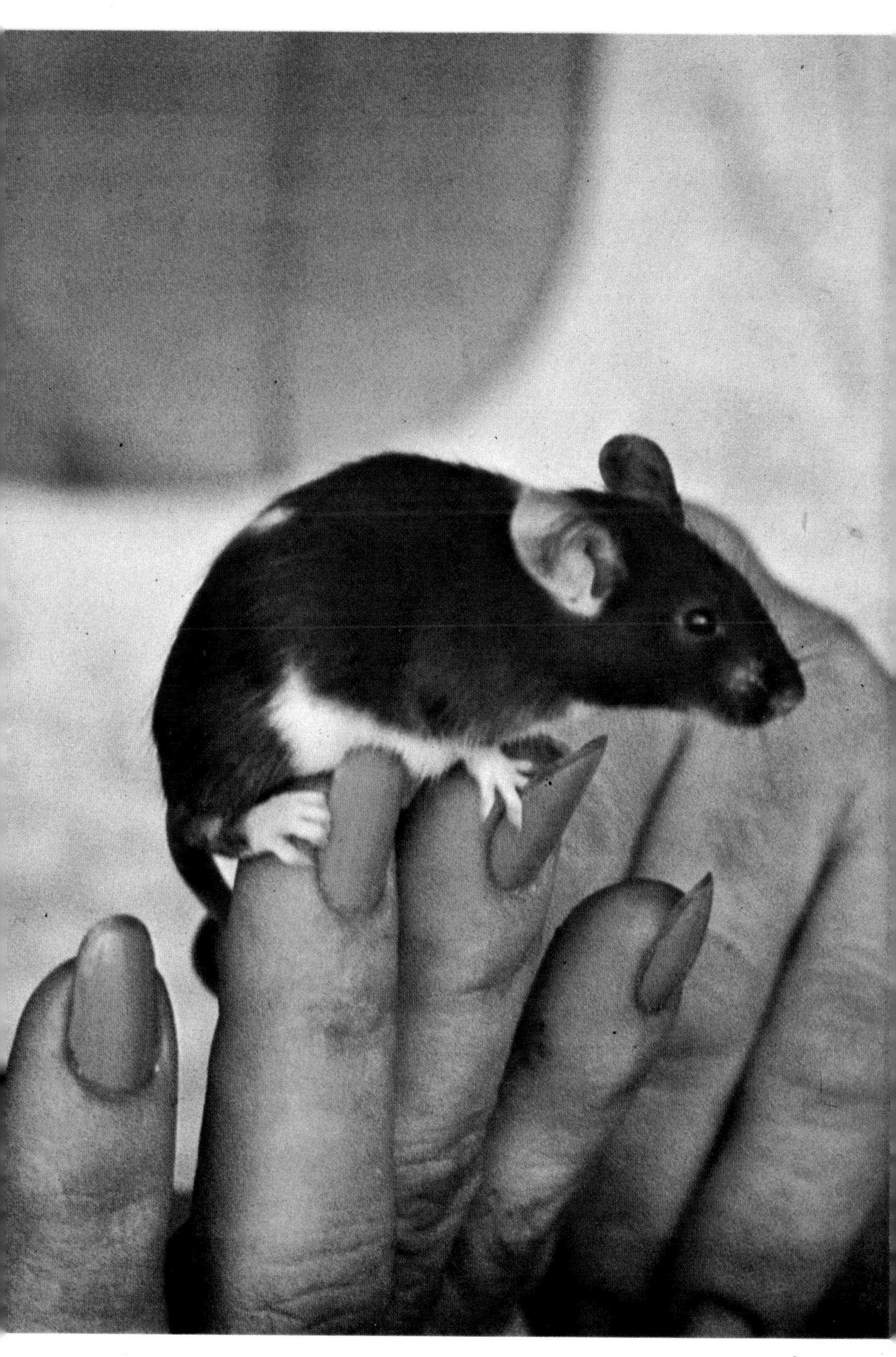

"The love of mice for music is quite peculiar. Harmonius notes entice them out of their hiding places, robbing them of all fear. They show up in broad daylight in rooms where there is an open piano; they like to run around on the keyboard and on the strings inside in order to be able to indulge in their favorite pastime of music. More than one reputable person has reported mice which literally learn to sing, that is, to twitter so as to sound like the song of canaries or other tame birds."

Your pet mouse is a mammal—a furry, warm-blooded creature which suckles its young. Contrary to popular opinion, not all mice are destructive nor undesirable human company. Some are of immense help to farmers. Farmers distinguish between rodent friends and rodent enemies. Most mice fit nicely into the ecological balance of nature. Even some of the so-called "destroyer" mice have become excellent pets for the right master.

A whole new world of excitement opens for the child or adult who wants to keep a mouse for a pet. Also, the full-bloomed naturalist who wishes a micro-world of his own in his own home which he can fully observe and photograph at his leisure finds mice very appropriate. They are sometimes inaccessible in nature because of their nocturnal and obscure habits.

Generation after generation of mice can be born and grown in your own home.

Just where does a mouse fit into the scheme of nature? Mice are rodents. So are rats, and rats may look or act very much like mice. Dividing lines between the mouse and rat are not always readily apparent. Generally, a domestic mouse is smaller than a rat and does not have as long or as pointed a snout.

And then there is the confusion about rabbits and hares belonging or not belonging to the rodents. They do not.

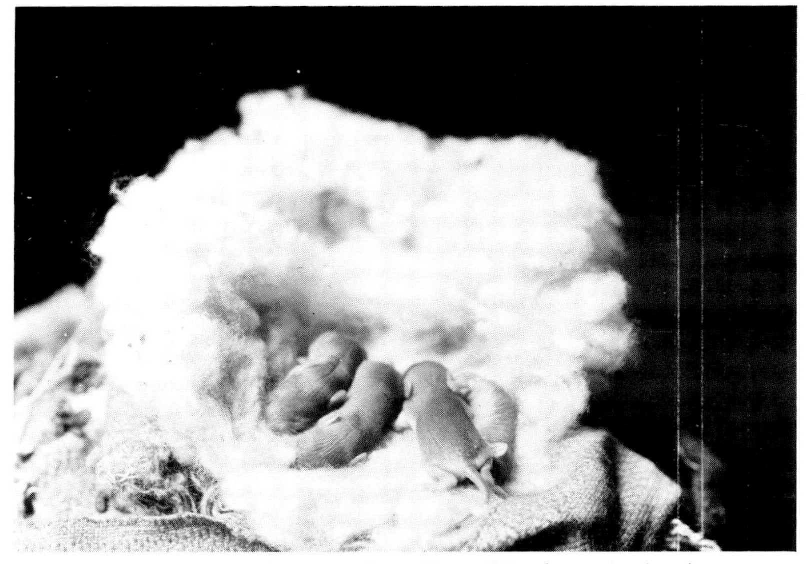

Only a few days old, these Canadian white-footed mice (see page 35), have had their nest opened for photographic purposes.

The jungles of the Congo (Zaire) produced two types of rats which are about twice as large as the Congo mouse which is the lower animal in the photo. Photos courtesy of the American Museum of Natural History.

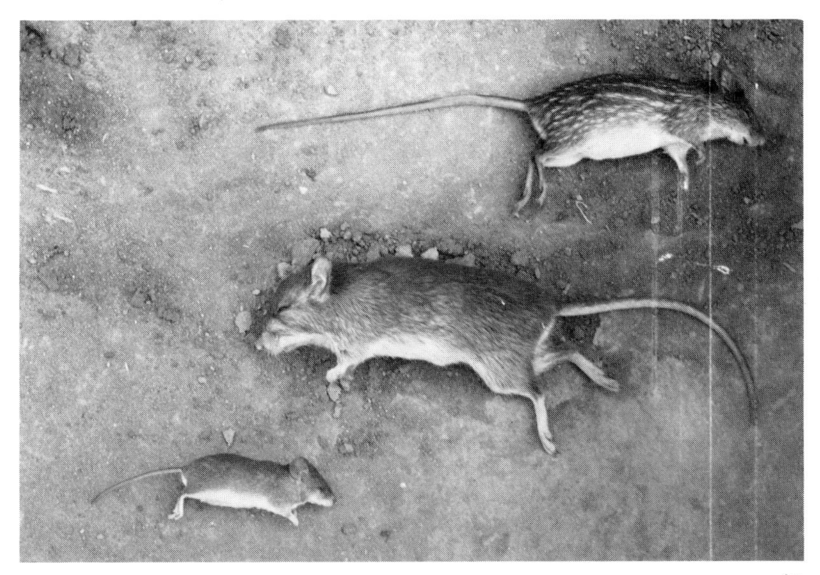

An albino common mouse with her naked, newborn young. If you reached for her babies she might bite you to protect them. On the facing page, pet rats photographed at a wholesaler's establishment. Photos by Claudia Watkins, Ferguson, Mo.

They used to be classified along with rats and mice, but are now considered separately as *lagomorphs.* Rabbits and hares, along with pikas, belong to the lagomorphs. In general, long ears, long hind legs and short, cotton-puff tail characterize rabbits and hares. They characteristically hop and leap (but the marsh rabbit may occasionally walk or run in a dog-like manner) and are vegetarians. Rodents have two incisors in the upper jaw, but lagomorphs have four incisors in the upper jaw. Lagomorph teeth are made for constant munching. Lagomorphs are herbivorous to the point of being serious agricultural pests when too many are at large. It may appear that a lagomorph has only two incisors just like the rodents, but if one looks carefully, one can see two large incisors in the rabbit's (or hare's) upper jaw, and two additional, smaller, incisors behind those front upper incisors.

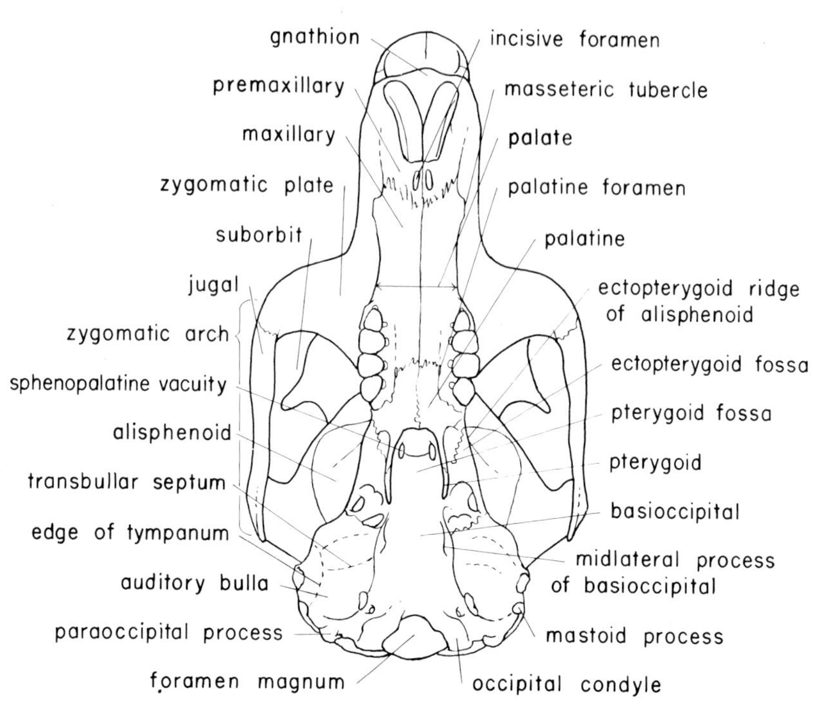

Palatal view (above) and top view (below) of rodent skull. Courtesy of the American Museum of Natural History.

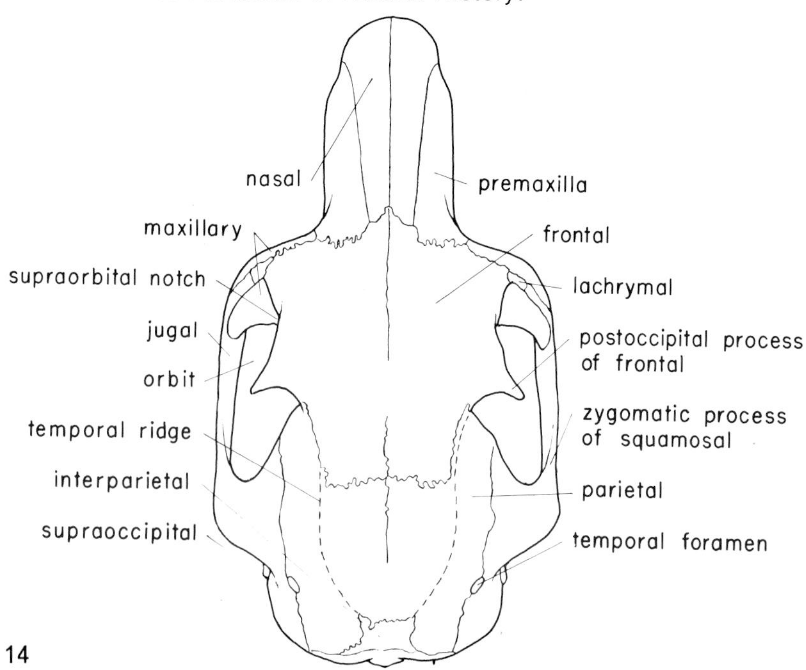

The Rodents or gnawers

The number of species in the rodent order exceeds the number of species in many other mammalian groups. At least 345 families of rodents account for 6,400 species, subspecies or races.

A common trait of the rodent is the presence of an upper and a lower pair of chisel-edged curved **incisors.** Three to five molars grow between the two incisors. The incisors grow all the time, their length being controlled by the animal's constant gnawing as well as the abrasive action of upper and lower teeth against each other. In addition to the sharpening action of incisor against incisor, this opposing action preserves the life of the rodent: if one incisor is lost or poorly positioned in the animal's mouth, the opposite incisor continues to grow, gradually arcing into the mouth until it pierces the roof of the mouth and thence into the brain. Or it may simply fill the mouth and keep the animal from eating, thus dooming it to a death by starvation. The incisors are separated from the rest of the mouth by a lipfold which protects a rodent from getting wood, metal, or any other substance into its mouth; that is, the incisors are used as a *scraping* tool—a plane—but do not necessarily act as *chewing* tools for the rodent.

Vibrissae, or tactile hairs, or just plain *whiskers*, are an essential part of animals built for "nosing around." They are generally found growing near the lips, eyes or cheeks of mammals. Even whales—our largest mammals—have hairs about the mouth, almost as if they were mere tokens of the whale's still being a mammal, like man, despite its aquatic habitat. Cats, notably, rely on tactile hair to make their pussyfooting way through obscure and jumbled passages at night. Bats were once thought to completely rely on such hairs to be able to avoid strings stretched across a darkened room (but now we know that the bat's "radar" signals also contribute).

Mice have been bred in many interesting colors. Their color has nothing to do with their ability to become friendly, but the more colorful they are, the more expensive.

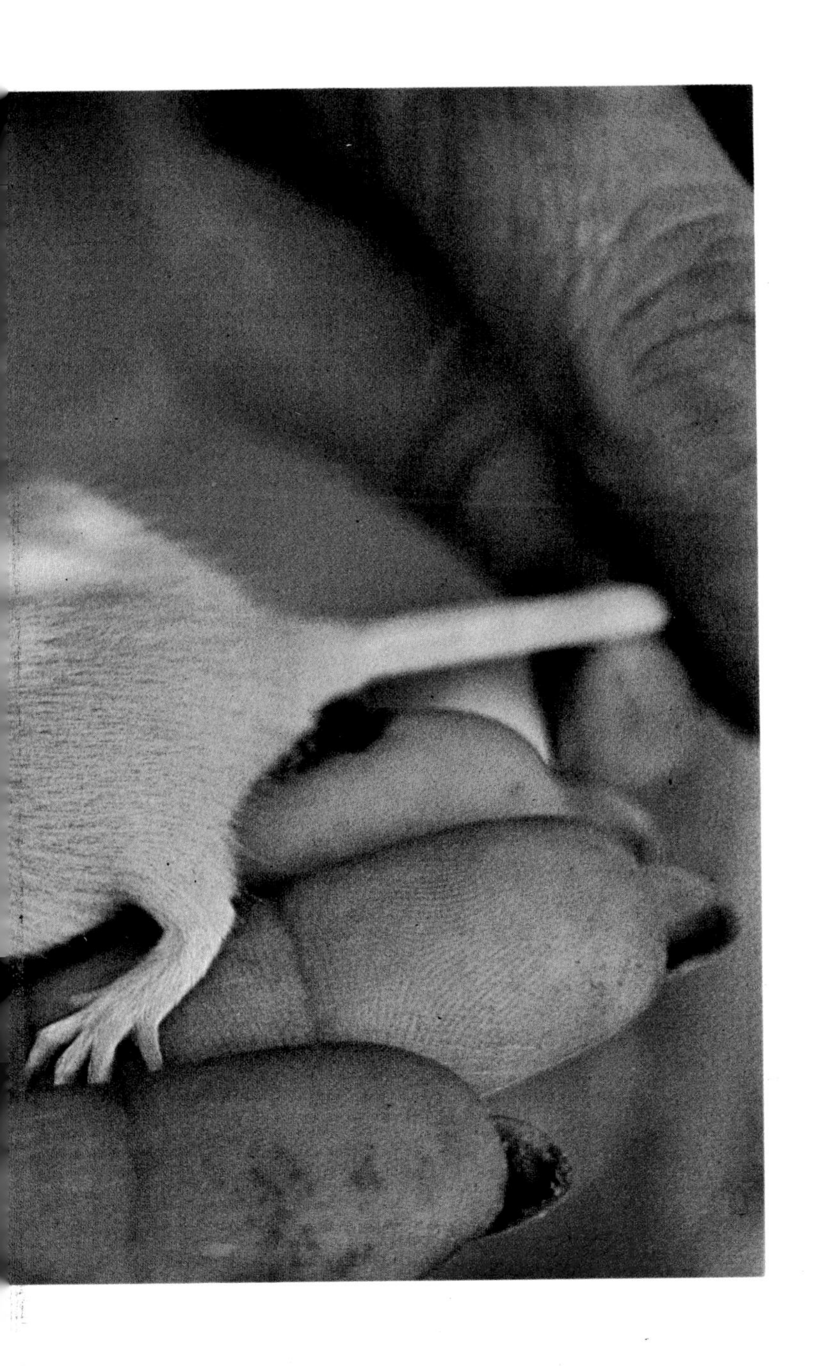

Studies have shown that the rat's vibrissae—when the animal is running in the open—extend out beyond the snout, where they evidently play a sort of distance-sensing role. The hairs which are placed most forward sweep the ground ahead of the running rat, the failure of the hair to detect any surface ahead perhaps warning of an impending fall. Side hairs on the snout "scan" edges and vertical surfaces as the rat scurries along. The rat's poor vision is thus in some measure compensated by these tactile hairs. (The cat, too, which also extensively relies on vibrissae, has poor vision as far as immobile objects are concerned, although movement quickly attracts its eye.)

Rodents are usually small, terrestrial animals, but there are burrowing, arboreal and aquatic ones, too. Most are vegetarians, but there are also carnivorous and omnivorous ones.

Rodents are universally distributed, being represented not only by mice and rats themselves, but also by squirrels (including flying ones), marmots, beavers, hamsters, gerbils, lemmings, voles (or field "mice"), dormouse, pacas, capybaras, goundi, agouti, coypu, rat-moles, porcupines, and guinea pigs.

Mice (and the rats scientifically classified along with them)

Mice and rats are the best suited of all the mammals to all situations, that is, of course, except for *Homo sapiens*. New world mice and rats adapt to quite a few habitats: rocky and mountainous ground, tropical forest areas, wide open spaces, water and aquatic areas. Aquatic areas include marshes, riverbanks, and flooded fields.

The following list of mice, rats and closely allied creatures will present an idea of the great diversity of the mouse and rat families.

(Above) A jumping mouse, *Zapus campestris*. (Below) A striped mouse from Zaire. Photos courtesy of the American Museum of Natural History.

Vibrissae (whiskers) are important for most rodents. On the facing page, the closeup of a pet hamster's face. Above we see a new color variety of hamster with its extended whiskers. The rat in the lower photo also has a face full of whiskers. Photos by Dr. Herbert R. Axelrod.

I. *Heteromyidae* family

Pocket mice, kangaroo mice, kangaroo rats.

II. *Cricetidae* family

Muskrats, leemings, voles, harvest mice, white-footed (or deer) mice, pygmy mice, grasshopper mice, woodrats, cotton rats, rice or field rats.

III. *Muridae* family

Black rats, Norway or brown rats, house mice

IV. *Zapodidae* (jumping mice)

The **Heteromyidae** have the following family characteristics: they burrow to build nests, are nocturnal, inhabit arid or semi-arid habitats (and do not need drinking water), are small, have fur-lined cheekpouches, weak forepaws but strong hind limbs; their tail is usually longer or equal to their body length; the front faces of the upper

An agouti, a rodent related to the guinea pig, found in Latin America. Photo courtesy of the American Museum of Natural History.

incisors are grooved (except for the Mexican pocket mouse); they are not destructive to man's activity (they are indeed beneficial in some cases because they eat weed seeds).

The **Cricetidae** family has the following characteristics: small to moderate in size (two to ten inch body length plus two-fifths to eight inches of tail, except for the fourteen-inch body of the muskrat); its members have five-toed hind feet (usually four-toed front feet but a few have five-toed front feet), long tails (except voles and lemmings), large eyes and ears (except voles and lemmings), varied habitats (aquatic, arboreal, rocks, terrestrial.)

The **Muridae** are colored a dull grey-brown to black. Long, hairless tails of uniform color are characteristic. The white laboratory mouse and rat are albino forms of the house mouse and the Norway or brown rat, respectively, and all belong to this family. In sheer numbers of individuals almost a quarter of all four-footed mammals in North America belong to the *Muridae*.

Wild forms of the *Muridae* are usually destructive. Laboratory specimens, however, are a valuable aid to mankind. As pets, some are satisfactory if kept properly. (Do not permit these albino forms of the wild Norway rat and house mouse to escape, for they will breed right back into the wild pest population!)

Members of the **Zapodidae** family are small to medium in size, have very long tails and large hind feet. The color of the body is yellow-orange on the flanks, darker on the back, and white on the abdomen. Buff or white edging trims comparatively small ears. Grooves run down the front surface of the upper incisors. Winter hibernation and a moist meadow or forest home are characteristic, too.

The following short descriptions of individual mice are given for a general orientation as to what to expect from wild or domestic mice as pets.

Other pet rodents include the gerbil (facing page), the guinea pigs shown above, and a chipmunk seen in the photo below. These and other rodent pets are available at most petshops. All can live in unheated areas.

The House mouse

The "standard" house creature since long ago, the house mouse has accompanied great waves of migrating peoples over the face of the earth. It reaches a length of three and one-half inches head and body plus two and one-fourth to three and one-half inches tail, and weighs one-half to one ounce. It does not hibernate, but remains active, with a respiration rate of between 1,530 to 3,500 breaths a minute, and a heartbeat of six hundred beats per minute (ranging from a low of 328 to a high of 780 in difficult situations). About five years has been reported to be the maximum lifespan of the house mouse, although some records report only two and one-half years as the maximum, and one to two years seems to be the usual lifespan. Reputable observers have reported that they have seen singing and dancing mice of this species; such "performing" individuals are probably ill mice.

Young house mice are slate-gray. Adults of both sexes show only slight seasonal variation. Both sexes are gray-

A common house mouse. Photo courtesy of the American Museum of Natural History.

brown or yellowish above, paler below, with gray-black hairs mixed in. Brown feet and dusky tail add the finishing touch.

House mice have no definite mating season, but breed all year round. Mice begin to mate at six to eight weeks of age, carry their young from nineteen to thirty-one days, and give birth to one to twelve (usually four to eight) pink and naked babies. They have four to eight litters a year. Young mice are weaned in three weeks.

These mice eat grain, vegetables, meat and just about anything from the table of their masters.

House mice normally live in open grassy areas when not keeping human beings company in their homes.

Jumping mice

Although similar in appearance to kangaroo rats, jumping mice are related to gerboas (kangaroo rat-like rodents). Jumping mice can spring incredible distances and are a pleasure to observe. In nature, they are rather difficult to catch, but once caught, are usually docile. These high bounders are slender and graceful creatures with hind legs and long tails—used to balance and steer when leaping, kangaroo fashion.

The meadow jumping mouse is usually nocturnal. This fellow can leap ten to twelve feet into the air (at three miles per hour in a vertical leap), swim, and burrow below the frost line. And he squeaks, too. He measures three to three and one-half inches plus four to five inches of tail, and weighs one-half to one ounce.

September to April is hibernating time for the meadow jumping mouse; he may even drop off to sleep during cold spells in the summer as well.

Meadow jumping mice mate in the spring when emerging from their hibernation dens deep under ground. Thirty days later a litter of two to six blind and hairless (but with whiskers!) babies are born. It would take thirty

A female climbing mouse, above. A jumping mouse, below. Photos courtesy of the American Museum of Natural History.

The young of most small mammals including voles, moles, opossum, and porcupines, make good and interesting pets. The adults cannot be trusted. Photos by Dr. Herbert R. Axelrod.

of those babies to weigh one ounce. They are furred in seventeen days, can see in three weeks, and are ready to strike out on their own in four to six weeks. Meadow jumping mice eat insects, shoots and roots, seeds and fruit.

The woodland jumping mouse, another of the jumping mice, can leap at ten to twelve miles per hour when frightened. It is four inches long, plus ten inches of tail. After half a year of hibernation, these mice mate and have litters of three to six babies twenty-one days later. They then have one more litter in September before hibernation season starts again.

Pocket mice

Pocket mice are great storers of food—nuts and tidbits—in many a crevice and hole. Do not confuse these gentle creatures with the pocket *gopher* (whose unpleasant nature makes it quite unsuitable for captivity.) The pocket mice resemble the gopher because they have cheek-pouches, but pocket mice are much more delicately built and colored.

It is thought that the pocket mouse goes without drinking water in its southwestern desert habitat. Water is believed to be derived from its food.

If you capture a pocket mouse by hand, it probably will not bite, but will squeak. Hold it gently awhile and it will calm down, not showing any fear. Later, if the pocket mouse is held in the hands quietly, or left undisturbed in one spot in its cage or on the ground, it will blink and close its eyes because of the light.

The pocket mouse is not an ideal mouse for persons who cannot stay awake at night to watch this nocturnal pet (who sleeps during the day). But there *is* a way . . . try observing your secretive, nocturnal pocket mouse in a room illuminated only with red light. The mouse will go about its scratching and scurrying around as if it were quite alone in deepest night time.

A prairie dog, so named because this rodent barks! Courtesy Catskill Game Farm.

The pocket mouse's tail is characteristically swollen in the middle. The silky-haired kangaroo mouse—whose head appears to be too heavy for its body—also has a swollen midportion of tail. That is, the base of the tail and the tip are of less diameter than the middle.

Grasshopper mice

The grasshopper mouse, Missouri mouse or mole mouse is man's ally in controlling such insect pests as scorpions, caterpillars, and other species of mice. The grasshopper mouse quickly becomes used to human company. It is not as timid as other mice and may even like to be petted when captured (although a few individuals may bite once caught.)

A meadow mouse. Photo courtesy of the American Museum of Natural History.

The pika or cony. Photo courtesy of the American Museum of Natural History.

Grasshopper mice are not always recommended as children's pets, but are quite capable of cleaning your kitchen or cellar of cockroaches. To do this, place the cage of grasshopper mice—the door opened—in the room to be policed, and let your grasshopper mice have the run of the kitchen or other room for the night. By morning, they should all be back snuggly in their nests, and the room free of cockroaches.

A Congo *Thrynomys* from Zaire (above) and a beautiful porcupine (below). Photos courtesy of the American Museum of Natural History.

Grasshopper mice are nocturnal animals of prairie and southwestern U.S.A. desert. Their coloration betrays their habitat: their short fur is gray or pinkish cinnamon above, white below. And the short (two inch or so) white-tipped tail at the end of a stout (two and one-half inch) body certainly identifies them in the field.

Young grasshopper mice are "mouse gray" above and white below. Adults of both sexes are identical in coloration, showing some seasonal variation. They are gray-brown above, darker on the back, lightening to yellow-red on the flanks. White highlights the feet and outer parts of the forelegs. The tail is brown-black above, white below and at the tip.

Grasshopper mice do not hibernate but are active year-round. After mating, females give birth to a litter of two to six babies, thirty-two days later. Fur grows on the babies by day twelve, their eyes open by day nineteen, and the females first mate by their third month of life.

White-footed mouse, or deer mouse

Pure white feet and underparts account for one of this mouse's names, and its fawn or brownish-grayish back color accounts for its other name of deer mouse.

The white-footed mouse does not hibernate, even in northern latitudes. It does keep out of sight, though, and lives as a nocturnal animal in its forest home, where it feeds on dried seeds, berries, nuts, acorns, grain, dead birds, and other dried meat. White-footed mice have been found nesting in buffalo carcases, right *in* their food supply!

Easily attracted to human beings, white-footed mice will soon eat from your hand, yet they are aggressive among themselves. They may fight one another. (Do not keep too many in one cage.) Mad dashes in the cage burn up energy otherwise expended in traveling through their forest habitats. In these dashes, their heartbeat probably exceeds their normal maximum of 858 beats per minute.

A white-footed mouse. Photo courtesy of the American Museum of Natural History.

And they sing as do house mice! What we hear, however, is probably only the audible (to us) part of a whole range of inaudible twittering communication.

This medium-sized mouse (three and three-fifths inch to four and one-half inch head and body plus two and two-fifths to four inch tail) has several litters yearly of one to nine young. The male helps in nest building, leaving its defenses, however, to the female. After mating in spring, the female gives birth to wrinkled, pinkish, deaf and blind babies who are transparent enough to let us see milk in their stomachs. Their eyes open by the second week (fourteen to eighteen days) and they are weaned by the twenty-first day. Females mature by the twenty-ninth day and males by the thirty-ninth day. Mating activity continues for almost two and one-half years but their lifespan is up to about five and one-half years.

Harvest mouse

The European harvest mouse (two inches plus tail long, one-sixth of an ounce in weight, reddish brown with white front)—the only European mammal with a prehensile tail—weaves a hanging nest of grass blades, suspended from small branches, plant stems or cornstalks. When the weather becomes colder, the summer nest is abandoned for a warmer, underground nest. Here, the mouse torpidly passes the winter. In addition to its prehensile tail and its summer home construction, the European harvest mouse is also a good swimmer. Both sexes of the adult harvest mouse are identical in coloration, and vary somewhat according to season.

A house mouse. This is a Dutch albino. Photo by Harry Lacey.

An English dormouse. Photo courtesy of the American Museum of Natural History.

Spiny mouse

The old world spiny mouse, a desert creature, is somewhat larger than a house mouse, and looks like a miniature hedgehog because of the short spines covering its back.

Spiny mice can be fed grains, seeds, green vegetables, and fortified dog food. The spiny pocket mouse of Southern California, according to analysis of its mouth contents, eats cantaloupe seeds, juniper berries, millet, corn, and peas.

A field mouse. Photo courtesy of the American Museum of Natural History.

A cony or pika from Colorado. Photo courtesy of the American Museum of Natural History.

A field mouse (pine vole) courtesy of the U.S. Bureau of Sport Fisheries and Wildlife. The photo below shows different colors of pet mice. Photo by Harry Lacey.

Other mice and mouse-like creatures of note

Our smallest mouse is a pygmy mouse, measuring about two inches from nose to hind quarters, plus one and two-fifths to one and four-fifths inch of tail, and weighing a quarter to a third of an ounce when fully grown.

The silky pocket mouse—or least pocket mouse—is only one and seven-eighths inches long (head and body) but its tail adds another two and three-quarter inches, thus making its total length longer than the pygmy mouse.

Long-tailed tree mice are nocturnal tree dwellers who nibble on the tender conifer needles at the top of the forest.

"Earth mice" are gophers, not mice. Gophers have yellow teeth which are visible in front even when the mouth is closed.

The gopher mouse or Florida mouse is a mouse, not a gopher.

Mention should be made of the field mouse or meadow mouse which is really a **vole.** Its four to ten-inch length is prolonged only by a three and one-quarter to three and one-half inch tail. A lifespan of only one and one-half years is understandable when one realizes that the vole has up to thirteen to seventeen litters of four to eight young per year, and that it must eat about its own weight in food daily to keep up such an active pace. There is no mammal which is more prolific. Meadow mice need a corner in which to "get away from it all" and some supplies with which to build nests.

Another vole—the pine "mouse"—is also characterized by having a much shorter tail in proportion to its own body weight: six inches head and body plus only a one inch tail. This burrowing, rock-dwelling creature has fur which can be petted smooth either forwards or backwards (associated with its burrowing habit: it can move easily in either direction underground) and is capable of broad jumps over crevices in its rocky habitat because of the increased traction provided by its hairy soles.

Voles make better pets than the sometimes cannibalistic **shrews.** As pets or zoo animals, voles have been fed commercially prepared mouse pellets, canned or dry dog food with vitamin supplements, canary seed, rolled oats, fresh green vegetables and other raw vegetables, as well as different kinds of fruit.

Shrews are insectivorous, mouse-sized mammals with beady eyes, semi-concealed ears and a very pointed, almost trunk-like snout. A shrew's teeth may be pigmented, and its feet are five-toed (most mice have only four toes on the front feet.) A shrew will starve to death in a very short time if it has to fast very long.

There are terrestrial and aquatic species of shrew. On land or on water their ceaseless predation for food will make them bold enough to eat from your hand (if they do not attack it as they would any other prey.) Tameness, however, is not to be expected from shrews. Be careful of making your finger appear too much like living game. And do not put two shrews together unless the cage is large enough to hold enough vegetation and rocks for the shrews to hide from (or stalk) each other.

You may eventually decide to release your shrews so that they can hunt their own food (which becomes quite a task to supply after a while). Meanwhile, feed them continually with live centipedes, spiders, worms, or even fresh meat from the kitchen. Also, fresh water and cereal such as oatmeal are good for the shrew in addition to the other, live foods.

Moles are extremely fine-furred (moleskin!) subterranean mammals with poorly developed eyes and powerfully built forefeet for clawing their way through the earth.

Johnson's Natural History (1867) reveals to us the nature of the mole:

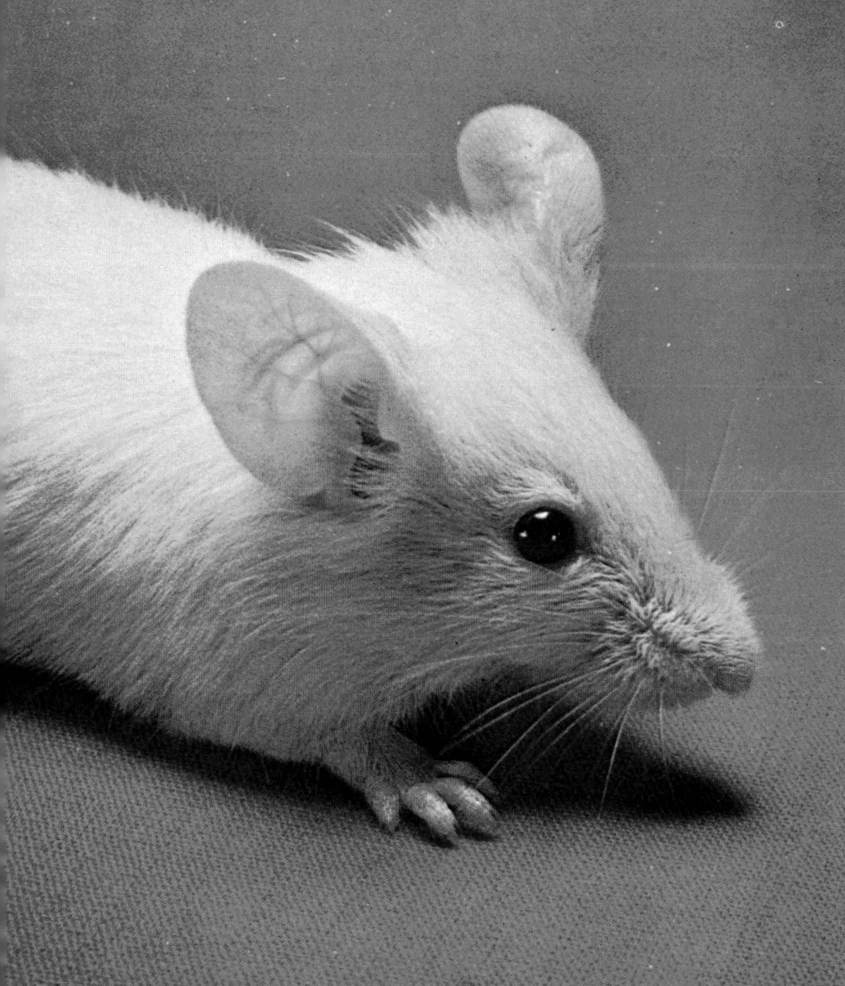

Young, but mature, this albino house mouse (and the brown pet rat on the facing page) is typical of the healthy animals to be found at your local petshop. Photos by Dr. Herbert R. Axelrod.

"We see a little, energetic, skillful miner, endowed by nature with all the tools needful for success in life. He seems condemned to toil and darkness, but if, as sometimes happens, he chooses to peep out from his burrow, and to take a night-scamper over the sod, his little eyes dilate and give him all the vision that he needs or wishes. Truly viewed, the mole, apparently condemned to a dark and dirty existence, is a happy example of a thrifty and contented housekeeper and a very model of personal cleanliness."

The American naturalist Thoreau had a fairly objective comment on moles: *"The moles nest in my cellar, nibbling every third potato . . ."*

Lemmings live in the far north, have small ears hidden by long and soft fur. Their tails are generally under one inch long. Alaskan Eskimo children are fond of making doll clothing from species of lemmings whose thick, white fur is typical of Arctic animals. The long-tailed lemming mouse is an arboreal species which prefers the Douglas fir tree.

And there are other creatures which have been named "mouse" because of their size or way of life. The so-called mouse opossum, for example, is an almost mouse-sized (but really closer to rat-sized) opossum, not a mouse.

TABLE OF COMPARATIVE SIZES

The measurements given in the following comparative table of sizes (made from statistical counts) are average, and individuals may vary smaller or larger. Other mouse-like or rat-like creatures are included for comparison. Note that the smallest of rats (there is one given at the end of the table for comparison) is also called a mouse: the rice rat, or rice field *mouse* or marsh *mouse.* This points out that the difference between mouse and rat, at times, may only be a difference in size.

Common Names	Length in Inches			
	Total	*Tail*	*Hindfoot*	*Ear*
House mouse	6.3	3.2	0.7	0.6
Grasshopper mouse, scorpion mouse mole mouse	6.0	1.7	0.9	—
Harvest mouse	5.0	2.2	0.7	0.4
Deer mouse, wood mouse, Eastern white-footed mouse	6.5	3.0	0.8	—
Lemming mouse	5.0	0.75	0.8	—
Long-tailed lemming mouse	6.0	2.4	0.8	0.2
True lemming	5.0	0.7	0.9 (hairy soles)	tiny and hidden in fur
False lemming, white lemming, snow lemming, Hudson Bay lemming, pied lemming	5.5	0.6	0.8	—
Field mouse, field vole, Eastern vole, common meadow mouse	6.7	2.0	0.8	—
Red-backed mouse, red-backed vole	5.2	1.4	0.7	—

Petshops usually keep their mice together for ease of care. The fact that albinos are kept in with normally colored mice indicates that your newly purchased albino female might give birth to normally colored babies!

Common Names	Length in Inches			
	Total	*Tail*	*Hindfoot*	*Ear*
Kangaroo mouse, jumping mouse	3.5-8	5-10	1.2	—
Pocket mouse	5.5	2.5	0.7	—
Pocket gopher, prairie gopher, red pocket gopher	11.0	3.2	1.4	—
Rice rat, rice field mouse, marsh mouse	8.8	4.4	1.0	0.6

This is the least desirable way to pick up a mouse.

HANDLING YOUR MOUSE

Everyone says "Don't pick up your mouse by its tail!" Good advice—the tail might skin off, if you grab a fleeing mouse by its tail. However, there *are* times when that is all you can seize . . . but let it go as soon as possible to release the strain on that organ (and it is an organ of balance.)

The figure illustrates a way of picking up an untamed pet. This is a method used in laboratories where mice are bred. A way to get tamed mice to begin with is to *see* your pet dealer pick them up first, and don't buy one unless he does!

Remember that pet mice usually bite from fear, not anger or vexation. Regular handling makes your new pet docile. To train your pet to sit carefree in your hand, set it on your palm, the forefinger and thumb of your other hand gently holding the base of the tail.

Restrain a mouse by holding it gently at the base of the tail. This is an off-color, dusty grey Siamese.

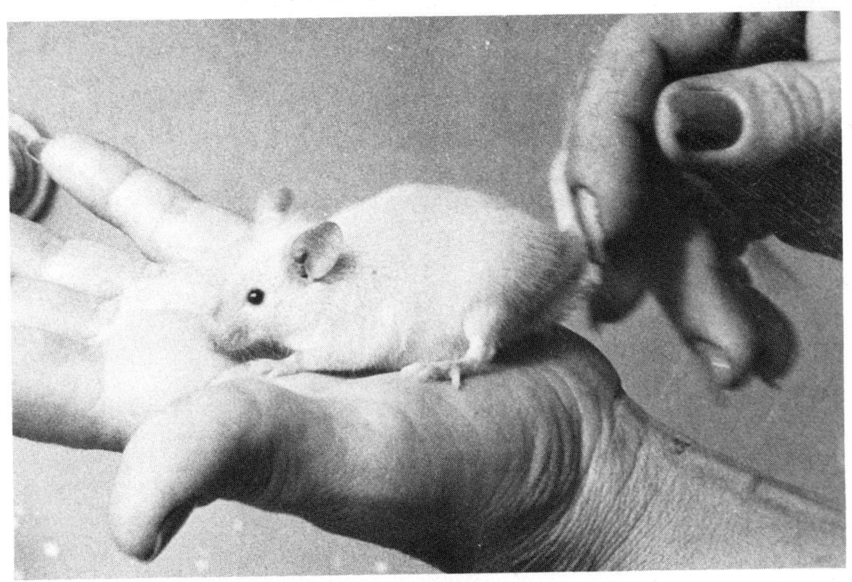

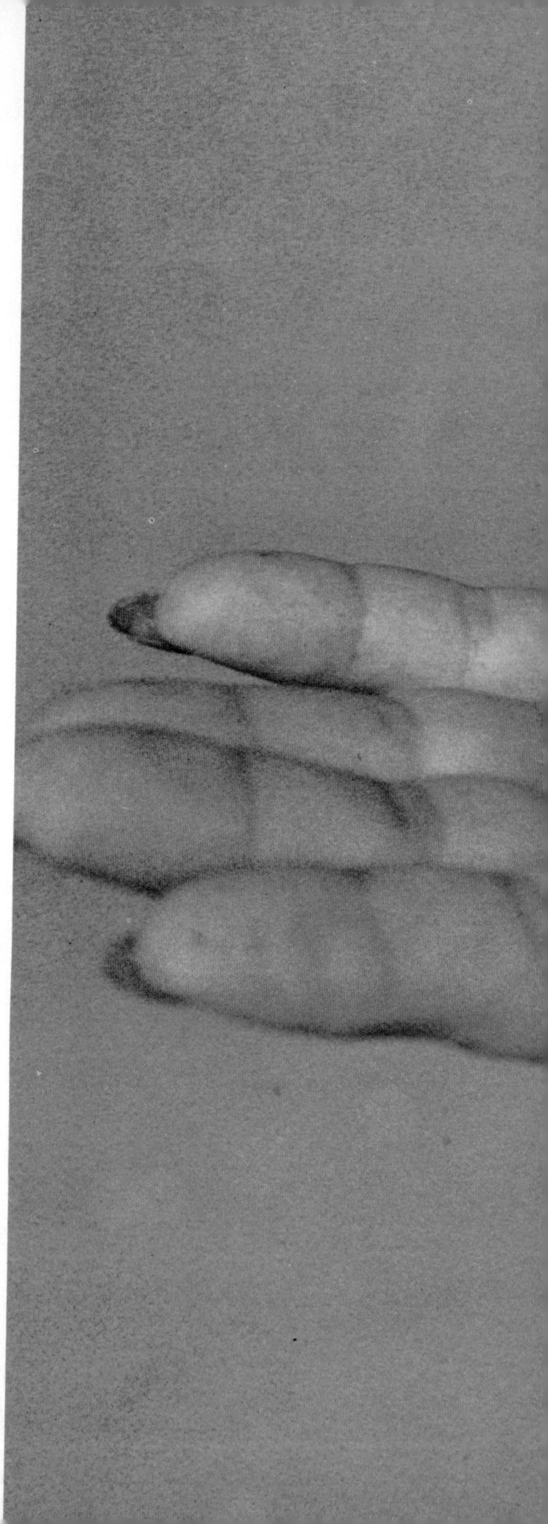

There is only one way to make mice tame. . . and that is to handle them. There is no substitute for handling and if you can get your pet to eat while it is in your hand then you can be certain the mouse is relaxed and you have a tame pet.

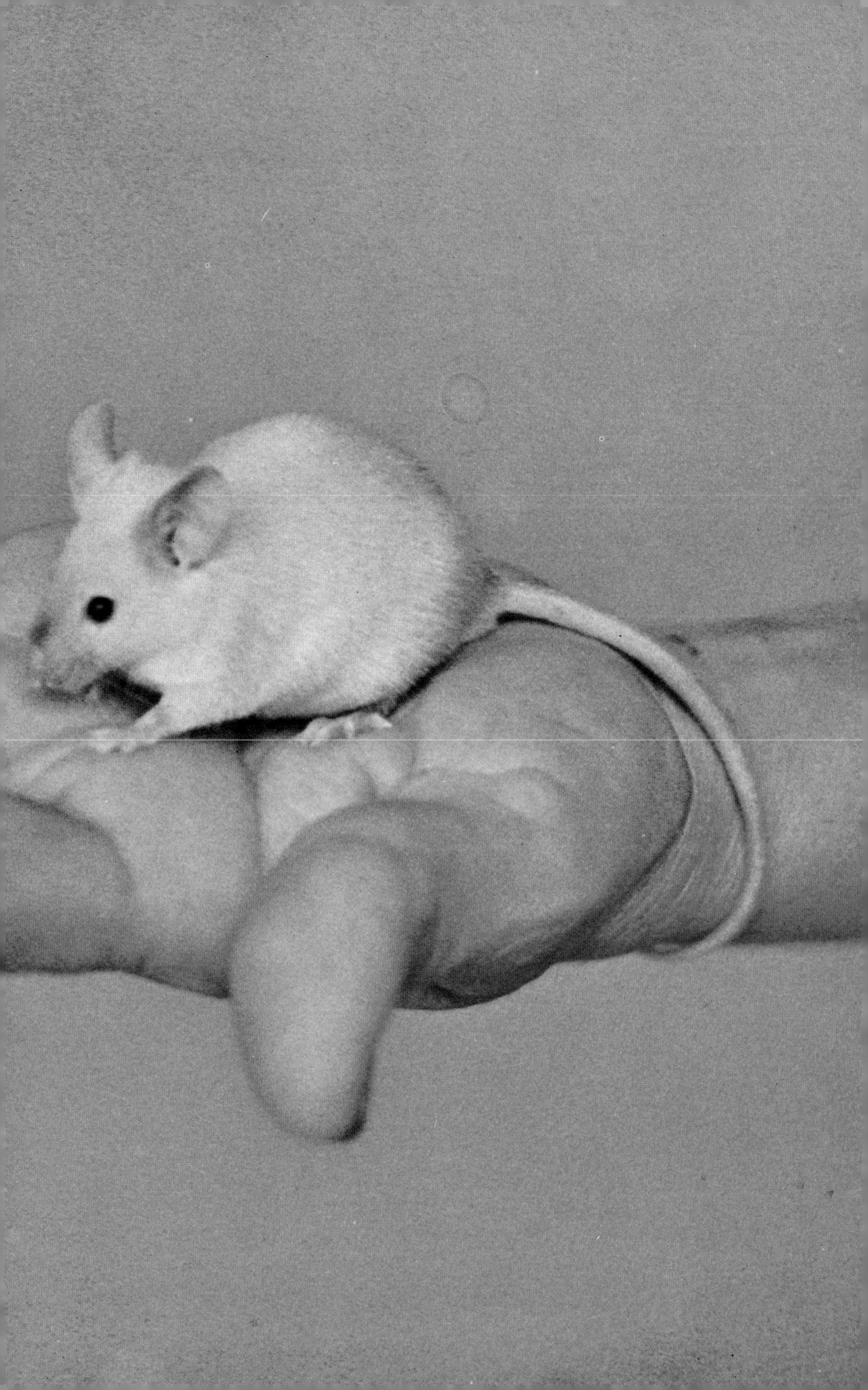

Handle mice as you clean their cage; this helps to tame them. Play with them at least two to three times weekly. Animals are *kept* tame, otherwise they may revert instinctively to untamed behavior. Let the mice know who is master. Be kind, yet firm and do not show your fear, if you have any. Mice catch on quickly. Associate feeding time with handling and petting. Your mouse will then look forward to eating and to your scratching its head while it eats. And, you can make the same noise (whistle, tch-tch, clicking tongue, kissing sound, raspberry, etc.) each

This is the best way to pick up a mouse, but don't carry him this way since he might become frightened and jump out of your hand and get hurt or escape.

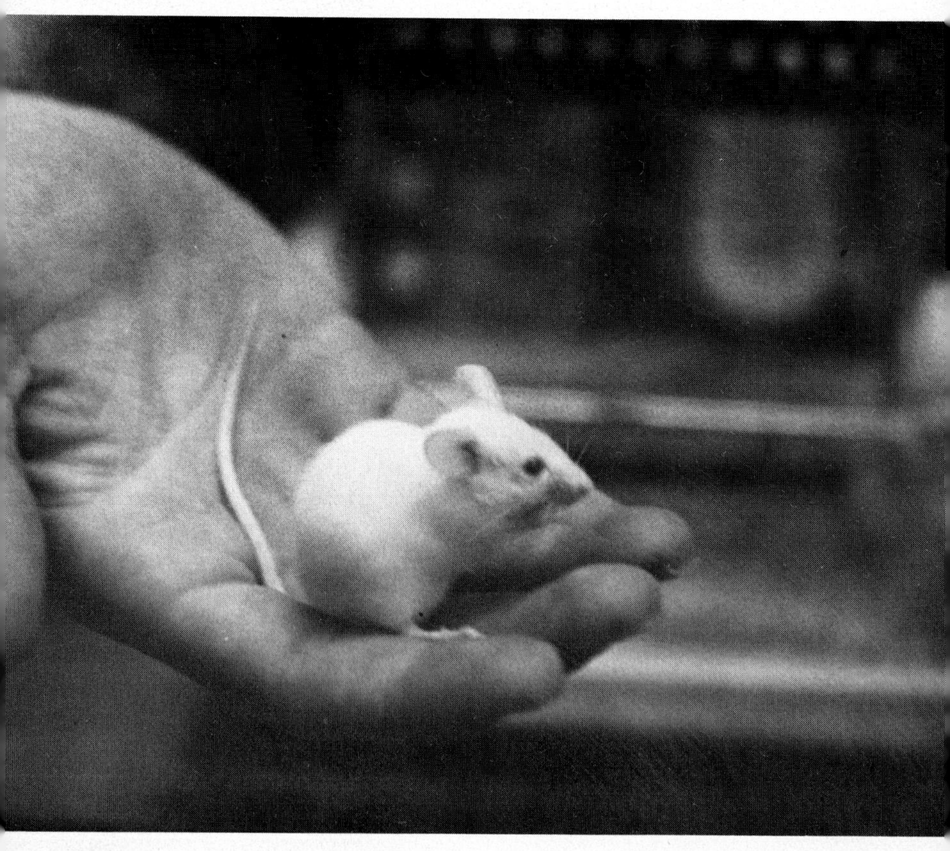

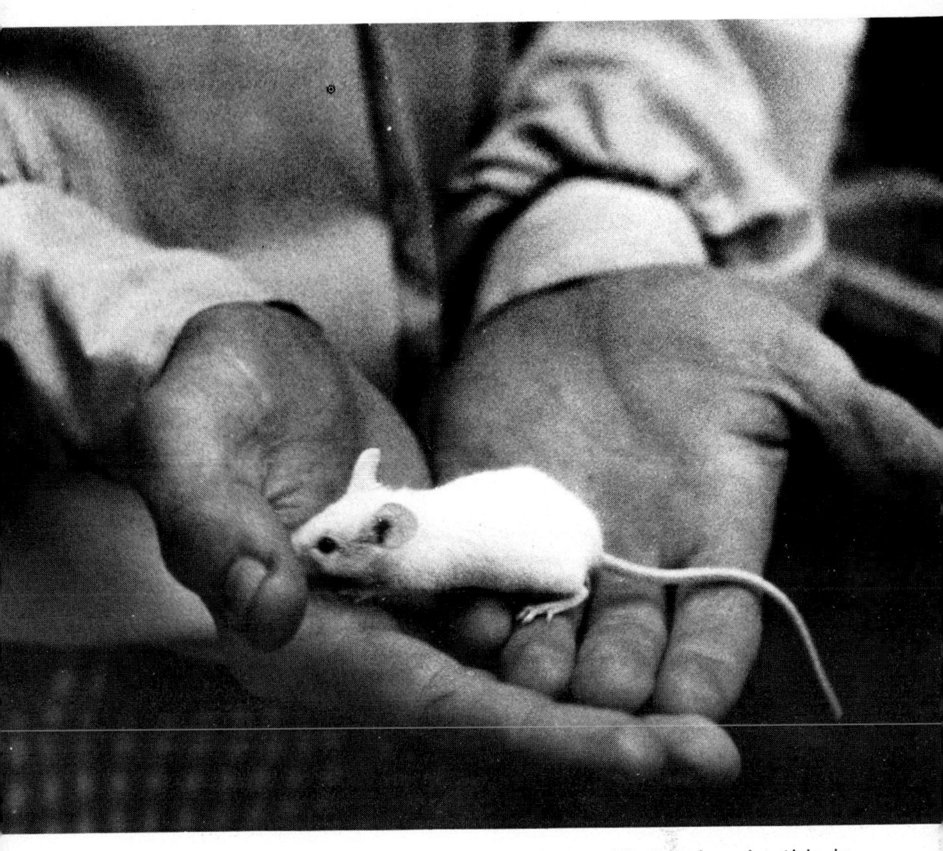

You can more safely carry your pet mouse with two hands; this is also a wonderful way for you and your pet mouse to become more friendly. You cannot handle a pet mouse too much. The more you handle it the more of a pet it becomes.

time you feed your mouse. Your pet will soon come to associate the sound with you and with food. Hand-feed rather than leave food in the cage, except when you are away overnight or even several hours (remember, rodents keep on eating, so leave enough food when you are away.) Mice which are frequently handled grow better and faster than neglected mice.

Fear quickly spreads among all mice in a cage. Wild mice panic easier than domestic mice. Make no sudden

Handling your pet mouse may be facilitated by grasping its tail and sustaining it in the palm of your hand. On the facing page, this is the proper way to walk with your pet mouse. Hold the tail with one hand and loosely clasp the mouse with the other.

movement. Do not make any slow movement which the mouse can instinctively interpret as the approach of a predator.

Mice can learn to walk upright (for only short distances, of course) by enticing them to reach upwards for a nut.

Although a mouse is just somewhat less intelligent than a rat, it is nevertheless a very sensitive creature. Lesson number one is: food is the reward. Your mouse will soon learn to make a straight line for your shirt pocket if it knows you keep peanuts there. A mouse will also learn to climb toy ladders, go in and out little doorways, and to do many other tricks.

MOUSE FOOD

Mice are easy to feed. Besides feasting on your table remnants, they can also be fed grain, birdseed, whole corn (you will certainly enjoy watching your tiny mouse hold a corn kernel between its paws and nibble on it like a squirrel), bread scraps, meat, chick-rearing pellets, seeding grass, cheese (of course!), fat, bacon rind, drippings, lard, fruit, diced carrots, linseed (for glossy coat), cod-liver oil for extra vitamins, bone (for keeping the teeth worn down short), and hard doggie bones.

Do not feed too much of anything at once; it becomes wet and putrid, forming a fine culture medium for the growth of bacteria and fungus.

Here is a mouse food test you may wish to use: present the mouse with a series of small jam, herring or peanut-butter jars, each one filled with a different kind of food. And put water in one of them. Each day you will be able to see how much of each kind of food is eaten and this will be the perfect taste test. The thing to watch for is which one is eaten up in one day with the least waste.

Young mice will eat bread soaked with milk warmed to body heat.

Hoppers or baskets designed to contain food pellets are good; mice, however, are apt to climb up onto the food pile and contaminate it with droppings.

Mice drink about a quarter of an ounce of water daily. Gravity-operated "demand" water bottles are available which allow the animal to serve itself.

Demand (or gravity-feed) water bottles, not troughs or pans, are the best way of providing a fresh, clean source of drinking water. Open containers of water get stepped in, overturned, excreted in, and can become an unsightly (if not downright unhygienic) mess.

Mice eat everything. . . meat, vegetables, insects, fruits and grains. Grains are preferred since they stay fresh and clean until consumed. Photo by Harry Lacey.

The plastic lab cage above has a week's supply of water in the demand bottle on the top. The wholesaler's cage on the facing page has three large demand type bottles which hold a few day's supply of water. Photo by Claudia Watkins, Ferguson, Mo.

A gravity-feed water bottle delivers water by gravity to a perforated cone or drinking tube attached to the neck of the bottle. The animal sucks or licks water from the opening in the cone or tube, and air bubbles enter the bottle to a place where the water is being drunk. No air enters and no water drops out when the animal does not lick or suck. The bottle will leak if there is a defect or incorrect adjustment in the system. Leakage can be caused by the wrong size hole in the cone or tube or a bad fit between the tube or cone stopper, or between the stopper and the bottle. Leakage is also caused by agitation of the bottle (water bottles cannot be used in moving vehicles.) Temperature fluctuations will alternately expand and contract the air in the bottle, pushing the water out. Contact between the hole which delivers the water and objects (or if the animal brushes against it) may

A typical rodent cage, suitable for hamsters, mice, rats or gerbils and the demand type of water bottle being affixed by the girl. Photo by Louise Van der Meid.

start the bottle emptying due to the siphon effect. Fungus, alga, calcarious deposits, corrosion, food or bedding might block the delivery tube or hole.

Bottles that are too small need to be filled too frequently. Bottles that are too large may empty themselves by siphon effect when they are only half full. When an animal drinks from a bottle, however, it will still be able to foul the water with saliva and the organisms contained in it. And some organisms multiply quite rapidly in ordinary faucet water. When room temperatures are high, such water may harbor more organisms than it did when you first put it in the bottle. So change it often.

An "ordinary" adult mouse eats about a teaspoonful of oats and a teaspoonful of moistened bread daily; nursing does usually require more food than this. (A nursing or lactating doe can eat her own weight in dry food in two days.)

Mice eat about five grams daily of the ordinary pellet foods on the market. If a good food pellet is given, no supplementary food is really needed. When we say "really needed" we mean for satisfactory growth. Supplementary feeding will increase the growth and breeding capacity of your mice to more than just a "satisfactory" level. This supplementary feeding can include vegetables, raw liver, cod-liver oil, milk, dry yeast, or many other things.

Good quality commercially prepared diets for specialized feeding purposes can be obtained from manufacturers. Special diets, too, are available for breeding. If you buy commercially prepared rations, do not keep them too long before use. Nutritional quality of stored dried foods decreases rapidly with time. Prepared foods, ideally, should be used within four to six weeks after compounding or preparation.

You should be aware that certain additives (such as estrogens) in prepared foods could affect breeding behavior and results.

Once a daily ration has been decided upon, stick to it at least for awhile; abrupt changes may disturb your pets. Enteritis and diarrhea may occur in new arrivals because of such sudden dietary changes. Gradually make changes in the diet of your new arrivals, if indeed you know what they were receiving before you obtained them. Do not overfeed your new arrivals.

The following dietary list includes foods known to be eaten by mice and similar creatures in the wild, as well as some of the foods which have been given to (and eaten by, of course) mice in captivity.

Common Names	*Foods*
House mouse	Omnivorous, preferring grain and vegetables
Grasshopper mouse, scorpion mouse, mole mouse	Insects, grasshoppers, scorpions, beetles, crickets, seeds, plants, and occasionally other mice
Harvest mouse	Seeds and grain
Deer mouse, wood mouse, Eastern white-footed mouse	Seeds, nuts, grains, acorns, berries, grass, leaves, dead birds
Lemming mouse	Roots, grass stems, green plants
Long-tailed lemming mouse	Green plants, insects
True lemming	Green plants, roots
False lemming, white lemming, snow lemming, Hudson Bay lemming, pied lemming	Green plants, grass stems, roots
Field mouse, field vole, Eastern vole, common meadow mouse	Omnivorous, mainly green plants, seeds, grass, grain
Red-backed mouse, red-backed vole	Omnivorous, mainly berries, seeds, tree and shrub roots and bark
Spiny mouse of California	Canteloupe seeds, juniper berries, millet, corn, peas
Shrews	Insects, live centipedes and spiders, worms, fresh meat, oatmeal
Kangaroo mouse, jumping mouse	Green plants, grass, seeds, berries, insects. Has been fed canary and sunflower seeds, apples, fortified dog food (canned), meal worms, rolled oats.

Mice need exercise if they are caged all the time and this exercise wheel made of stainless steel serves them well.

Pocket mouse	Grains, seeds, vegetable matter. Has lived seventeen years on canary or other mixed seeds only (no water, no green foods). Zoos have fed them canary or other small seeds, tiny portions of dry or canned dog food, vegetables. Water, though always available, is apparently never taken.
Pocket gopher, prairie gopher, red pocket gopher	Herbivorous : roots and green plants.
Domestic varieties of mouse	Whole oats, good hay, cubes of stale bread soaked in water. Occasional (several times weekly) birdseed, oats, dog biscuit (doggie bones), mixed birdseed, bread soaked in milk or water, fresh vegetables (some breeders do not use them—for best results, go by your own animals' preferences), hay, water.

Most often, colonies of mice are given staple diets of canary and sunflower seeds, oats, raw green vegetables, and fruits. Cod-liver oil is mixed in with dry or canned dog food. Sometimes meat and fish are added for proteins.

Meat-eating mice need to be kept supplied quite well with fresh meat (insects or whatever a particular species eats); occasionally, substitutes for fresh meat are accepted. Experimentation will yield much information on how each particular animal is willing (or indeed is even able to) adapt to "human food."

Vegetables should be raw, unwilted and washed clean of soil and debris. Remove uneaten vegetables immediately from the cage. Start feeding your mice vegetables gradually if they have not been given such food before, or diarrhea could be the result.

Hay (timothy, clover, etc.) is an important element in mouse diet. A fresh, clean odor—not a musty smell—characterizes the kind of hay you should use. Mice also enjoy tunneling through hay, and building nests in it.

Hard grain food, various grasses, and an occasional Brazil or other hardshelled nut will provide gnawing material for wearing down the mouse's constantly growing incisors. Acorns and beef bones, too, are suitable "teething" items. Bones, by the way, provide minerals as well as gnawing pleasure for your mice. Otherwise, these master gnawers will concentrate on gnawing through their cage.

At dusk your mice become most active, remaining awake until the wee hours of the morning. Feed them in the evening so that they will be nourished for their nocturnal activity. They can be trained to eat during the day, but will become hungry again at night. If it is inconvenient to feed your mice evenings, leave a cache of food for them in a container off to the side; this way, the mice will not contaminate it.

If the cage is cleaned out at least weekly, food may simply be placed on the floor of the cage; it should be eaten up within ten minutes or so.

To encourage young mice to fend for themselves, sprinkle oatmeal or other tiny pieces of grain food about the nest. This way, the little ones start to nibble at the earliest possible moment. When you breed for exhibition stock, every moment of growth helps. Fat growth, however, is undesirable. Too many "goodies" of the traditional mousetrap bait kind (cheese, bacon) make for fat (and unhealthy) mice.

A final word about diet: Vary the diet from time to time to experiment and find the best possible combination of foods discussed in the sections above, and also to provide for maximum lifespan, activity, size, and fertility.

MOUSE GROWTH CHART

(Average weight in ounces)

	Black piebald mouse		*White mouse*
	Male	*Female*	*(males only)*
Weight at Birth ...	0.05	0.05	—
Day 2	0.22	0.21	—
Week 3	0.28	0.29	0.29
Week 4	0.43	0.43	0.44
Week 6	0.67	0.65	0.69
Week 8	0.80	0.70	0.89
Week 12	0.91	0.83	0.89
Week 16	1.01	0.94	0.96
Week 20	1.06	1.02	0.98
Week 24	1.17	1.14	0.97
Week 28	—	—	0.99
Week 32	—	—	1.04

This chart provides only a general guideline. Mouse weight varies according to diet; sometimes a pet mouse fares better than his laboratory relatives (from whom the above weights were collected), and sometimes not. Also, weight may vary according to climate and health, or depend upon what species is being considered. Individual mice, too, show much variation, just as do people. A nervous mouse may get skinnier the more it jitters around . . . or it may get fatter from nervous eating, and so on.

HOUSING

Whether you need technical details for extensive mousery operations, or only a few notes on a shoebox-sized mouse house, perusal of this section should help you to come up with your own formula for a good mouse home. In general, keep it *warm*, *dry*, *clean*, and *safe* from cats, dogs and wild mice and rats.

Fancy cages at petshops are usually inexpensive, clean and easy to maintain, but you can make your own of wood. Wood is hard to keep clean and free of odors because the wood absorbs odors and urine.

Metal cages are usually used because mice can gnaw through quite a few substances. Take care that metal cages do not become too hot (near radiators) or too cold (in winter, outside, or too close to a window). Wooden cages may be suitable if reinforced with wire mesh.

In general, a cage for a doe plus her litter should be *about* fifteen inches long, six inches wide and six inches deep; for twenty to thirty "adolescent" mice, or up to a dozen adults, the cage should be *about* twenty-four inches long, eight inches wide and eight inches deep.

Recommendations for cage space have been drawn up by various authorities as follows:

Age	*Minimum space*	*Maximum animals*
Weaning mice ... (up to 5 weeks of age)	6 sq. in. per mouse	40 mice per cage
5 to 8 weeks	8 sq. in. per mouse	30 mice per cage
8 to 12 weeks ...	12 sq. in. per mouse	20 mice per cage
Over 12 weeks of age	15 sq. in. per mouse	20 mice per cage

The above figures are only general guidelines, it should be remembered.

HEALTH

Before buying your mouse, check to see whether the dealer's cages are clean. Does the mouse have a glossy coat? Are the mouse's eyes bright, its ears and teeth clean? Is the mouse plump but not fat? Does it move freely and uninhibited, and not jerk along nervously? Your healthy pet, if fed and housed well, will most likely do well unless it picks up an infection or is injured. Or, if its resistance is lowered due to poor food or environmental conditions, it may manifest one of the following: virus infection, bacterial infection, mite infestation, ringworm fungus (which causes bald spots in the fur), ringtail, or vitamin E deficiency.

A really great substitute for a cage is an old aquarium. . . even a "leaker". . . which can be closed in with a suitable mesh top. Photo by Claudia Watkins, Ferguson, Mo.

A bacterial infection would be salmonellosis or mouse typhoid. Mouse rheumatism or arthritis is also caused by bacterial infection, as also is mouse septicemia. Mousepox is a viral infection, as is infantile diarrhea. Mite infestation causes mange. Although your dealer or veterinarian may be consulted for a specific therapy, the main point is to avoid these conditions by choosing healthy stock to begin with, and maintaining them on a proper diet and under the proper environmental conditions.

Petshops are the best place to buy pet mice.

Aquarium tanks are fine if they are large enough; they permit an open view, are easy to clean, and are escape-proof (over ten inches high or fitted with a top which still allows air to enter). A layer of soil can be put in these waterproof aquarium tanks, too, thus making your

This is a perfect, inexpensive mouse cage. It is made of rustproof wire, has a suitable demand waterer, an exercise wheel and is easily cleaned.

mouse's home more attractive than a clinical cage would be.

If a metal dollhouse becomes your mouse's home, be certain to ventilate it adequately and provide for a moveable side, front or back, so that the inside can be cleaned properly.

Softwood shavings—especially cedar wood—make an excellent bedding and nesting substance. Sawdust, too, is very good, but it must be made from white softwood; hardwood sawdust may contain harmful natural chemical substances (phenols). Obtain the shavings or sawdust directly from the sawmill or pet shop. Otherwise, it could be contaminated with the droppings or hair of animals. Peat moss, although expensive in some areas, greatly minimizes odor. Its acid content counteracts the decomposition of animal droppings. Shredded paper is an old standby. Renew bedding once weekly whether it seems dirty or not. When you notice where your pets' favorite "toilet spot" is (they usually have one), pile up a little more sawdust (or whatever you are using) there, then change it daily.

A bottom layer of earth (under a top layer of sawdust, wood shavings, cat litter, etc.) is appreciated, for mice love to burrow; however, earth will cause wood to rot, so use earth only in cages with glass or plastic bottoms. Stainless steel, too, might be used, although metals in general may rust.

It goes without saying, of course, that your pet mice must be protected against predation. That means cats in particular. Poor Alice in Wonderland, however, did not realize it:

> *"Perhaps it doesn't understand English," thought Alice, "I daresay it's a French Mouse. . . ." So she began again:* "Où est ma chatte?" ("Where is my cat?") *which was the first sentence in her French lessonbook. The Mouse*

gave a sudden leap out of the water, and seemed to quiver all over with fright. "Oh I beg your pardon!" cried Alice hastily, afraid that she had hurt the poor animal's feelings. "I quite forgot you didn't like cats."
"Not like cats!" cried the Mouse in a shrill passionate voice. "Would you like cats if you were me?"

Colonies of mice are generally housed in wire-mesh or glass-sided cages with removable wire tops, or else in plastic bins fitted with wire tops, each of which holds an inverted, gravity-feed water bottle and a hopper to hold pellets or other food.

Mice, unlike rats, need one or more separate nesting sites in their cage. A nesting site can be a box, can, jar, flower-pot or other "cozy" container provided with a small opening for an entranceway.

Mice will build their own nests from any materials you put in their cage: paper (but without any printing on it!), clean hay, cloth and fabric fragments, for example.

Symptoms and control of diseases

(Note: Not all symptoms occur in each case of a given disease. Also, follow manufacturer's instructions on medications. Seek expert advice. Remember, treatment may be as dangerous as a disease itself. Good management and general hygienic measures may be the most effective care you can give.)

These mice are terribly overcrowded. Fortunately they will be out soon. Overcrowding means diseases are soon to come!

Disease	*Symptoms*	*Cause and Control*
Mouse typhoid	Rapid loss of condition, diarrhea, conjunctivitis (with pus). Mouse may die 1-2 weeks following infection, or may recover.	*Salmonella* bacteria is causative organism. Quarantine new arrivals. Disinfect housing.
Tyzzer's disease	Loss of condition, drop in weight, diarrhea.	*Bacillus piliformis.* Disinfect bedding. Avoid contamination with droppings from wild mice. Avoid overcrowding.
Mouse rheumatism or arthritis	Loss of condition, drop in weight, conjunctivitis (with some pus). Mouse might die a few days after infection; if mouse survives, legs, feet and perhaps the tail may swell with edema, and may ulcerate.	*Streptobacillus* (*Actinomyces*) sp. Avoid immediate introduction of new arrivals. Disinfect housing. Change bedding.
Mouse septicemia	Loss of condition, weight loss, difficult respiration, conjunctivitis.	*Klebsiella pneumonia, Corynebacterium kutscheri, Erysipelothrix muriseptica* and other bacterial species. Remove all mice. Disinfect cage.
Mouse pox (also called ectromelia)	At times death with no symptoms. Scabby lesion, skin eruption. Feet and tail may swell and become gangrenous.	A pox virus. Infected mice and contaminated objects spread the disease.
Diarrhea of newborn mice	Liquid, yellow diarrhea from mice under two weeks of age. Develop-	Modify diet. Take general hygienic measures. Genetic constitution of mice may

	ment of survivors may be retarded.	determine susceptibility.
Ringworm	Hair loss, scales, crusts.	*Trichophyton mentagrophytes* ringworm fungus. (This organism can infect human beings.) Take hygienic measures. Fungicides may be tried.
Body mange	Hair loss and perhaps inflammation of skin (usually in young mice and breeding females). Adult males may harbor the mites, but do not lose hair or have inflammation.	*Myocoptes muscelinus* mite. Dust with benzine hexachloride (Lorexane) twice weekly. Disinfect housing.
Head and neck mange	Inflammation, suppuration and scabs (usually in males).	*Myobia musculi* mite. Dip mouse in 1 :9 tetraethylthiuram monosulfide (Tetmosol) and warm water twice weekly. Or, dip in 0.2% solution of di-p-chlorophenyl methyl carbinol (D M C) in 50% alcohol.
Ear and body mange	Ears stopped up with a white waxy mass of mites. Mites also burrow under the skin and create pockets there.	*Psorergates simplex* mite. A large drop of dibutyl phthalate (keep out of mouse and human eyes!) in the ear; repeat in seven days if necessary. Do not try to dig out the waxy plug. For the body mange, apply a 2% aqueous solution of 15% 2-(p-tert-butyl phenoxy isopropyl-2-chlorethyl sulfate)—Aramide—twice a week.

Lice. Ear, body and head mites	See above.	Dip mice twice in a mixture of the following two solutions at a 1-hour interval, then transfer to clean housing. Repeat if necessary in 1-3 weeks: Solution A=2 grams DMC in 3 grams ethanol. Solution B=67 grams Tetmosol (25% solution) in 1 liter warm water. Mix solution A and B and use as indicated above at 37°C. (or 98.6°F.).
Roundworms and tapeworms	General condition may be affected; otherwise, no symptoms.	*Aspicularis tetraptera* and *Syphacia obvelata* roundworms and cystic stage of the cat tapeworm. General hygienic measures.
Poisoning by inhalation of chemical fumes or vapor	Death.	Mice are very sensitive to carbon tetrachloride and chloroform vapor.

These experimental mice are lethal-producing types. The roan (bottom), the yellow-red in the center, and the Danforth's short tail on top. These are not pet mice. They are strictly laboratory animals, specially bred for certain inherited characteristics, and produced by special breeders who supply only laboratories. Photo by Claudia Watkins, Ferguson, Mo. →

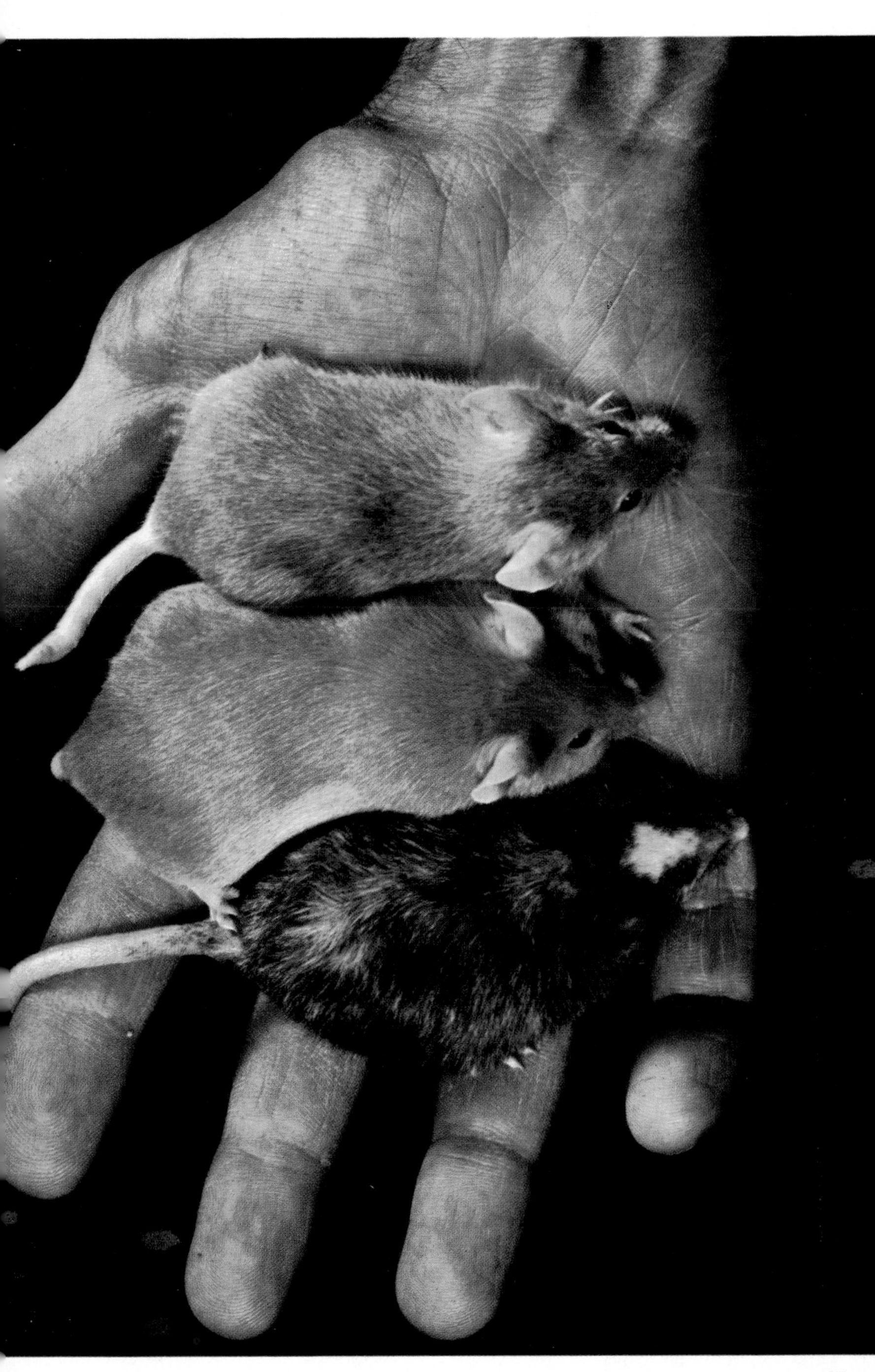

So much for sick mice. If *you* should get sick, however, remember that the Pennsylvania Dutch once recommended *mouse pie* for prevention of bedwetting and smallpox, among other ills. (Joke!)

Lifespan

Species	*Lifespan*
House mouse	Five years
Grasshopper mouse	Three and one-quarter years
White-footed mouse	Five years, eight months
Common dormouse	Three years, eleven months
Fat dormouse	Seven years
Spiny mouse	Four years, five months, twelve days
Jumping mouse	Six months
Pocket mouse	Seventeen years

Spread of disease

The spread of disease among the mice in a housing community of them depends upon the following factors:

(a) How many animals are already infected?

(b) How infective is the disease-causing organism? How soon do adequate numbers get from the first sick animal to the second sick animal, and so on?

(c) How potent is the disease-causing organism, and how well does it counter the animal's resistance?

(d) How resistant is the animal against a particular disease-carrying organism? (Some resistance is natural, and some is acquired by previous bouts with the disease, or by vaccination or inoculation.)

Remove the male from your mouse family scene if the diet is deficient in protein; in this case the male might eat the young.

Shallow ashtrays are good for milk to help the doe mouse with nursing, and will avoid the young mice drowning which they might do if the milk containers were any deeper. As bedding material in the cage, newspapers are excellent; strips of it are warm, absorb well, and can easily be destroyed.

Do not handle the naked, newborn mice until the doe brings them out of the nest, or she may eat them.

A note on treatment . . . and not treating

The treatments suggested here are some of the measures tried by others. Do not consider them sure-cures. Avoid treatment with drugs and medications unless you have expert advice, or clearly understand the instructions which come with the medication. In many instances, mousebreeders "dispose of" an ill mouse rather than try to cure it. With pets, of course, one does not wish to take such a drastic step. Consult a veterinarian—a Doctor of Veterinary Medicine, D.V.M.—when in doubt as to the proper, safe treatment.

Hygienic measures go quite far in preventing disease and curing it by allowing the mouse's own resistance to help. Nutritional support such as vitamins may help, too. Keep food and water fresh. Keep housing and bedding clean and dry. Remove accumulated droppings. Although animals should be fondled and played with to keep them tame (and they enjoy playing, too!), let them rest quietly awhile if they seem indisposed. You can still "sit by their bedside" and talk with them until they recover, but reduce any rough-house for awhile.

Terminal disinfection

Dead mice should be promptly removed from a group of mice, the other mice transferred, and the box cleaned and disinfected. This may stop contagious infections from spreading to healthy mice although some may already be infected: hygienic maintenance of the cage, however, will hold down the disease manifestations or perhaps limit the severity of any outbreak when and if disease finally appears after an incubation period.

BREEDING

Overbreeding—not underbreeding—will be the problem. A single pair of mice, once they reach two to three months of age, can produce about one hundred and thirty-five mouselings a year. Mice start early. There is no definite mating season. You should stay with a modest number of animals and keep in mind that good initial stock will increase your chance of selling surplus mice and doing well at mice shows. Color breeding is not complicated unless you go in for it as a specialty.

Provide an adequate selection of nesting sites. See the chapter on housing for more notes on nesting sites. The doe's expectant condition can be seen easily: her teats stand out prominently and her abdomen is swollen. Do not handle her too much . . . or at all. Give the expectant doe a private cage, if possible, which contains a nesting container. Provide her with nesting materials (frayed remnants of cloth or rope, paper, etc.). Extra rations of milk (from a gravity-feed water bottle) will be fully appreciated.

When you peek at the newborn mice, cull out the dead and malformed ones, but do not touch the other ones, or the doe.

Mating to weaning time and other vital statistics

The figures below are averages for several species *in general*, and are meant only to provide a notion of relative time. Each species shows quite a bit of variation. See under individual species for closer approximation.

Children love mice and they make excellent pets. More than 2,000,000 pet mice are sold each year and they are becoming more and more popular as they are introduced to children in their elementary school education. →

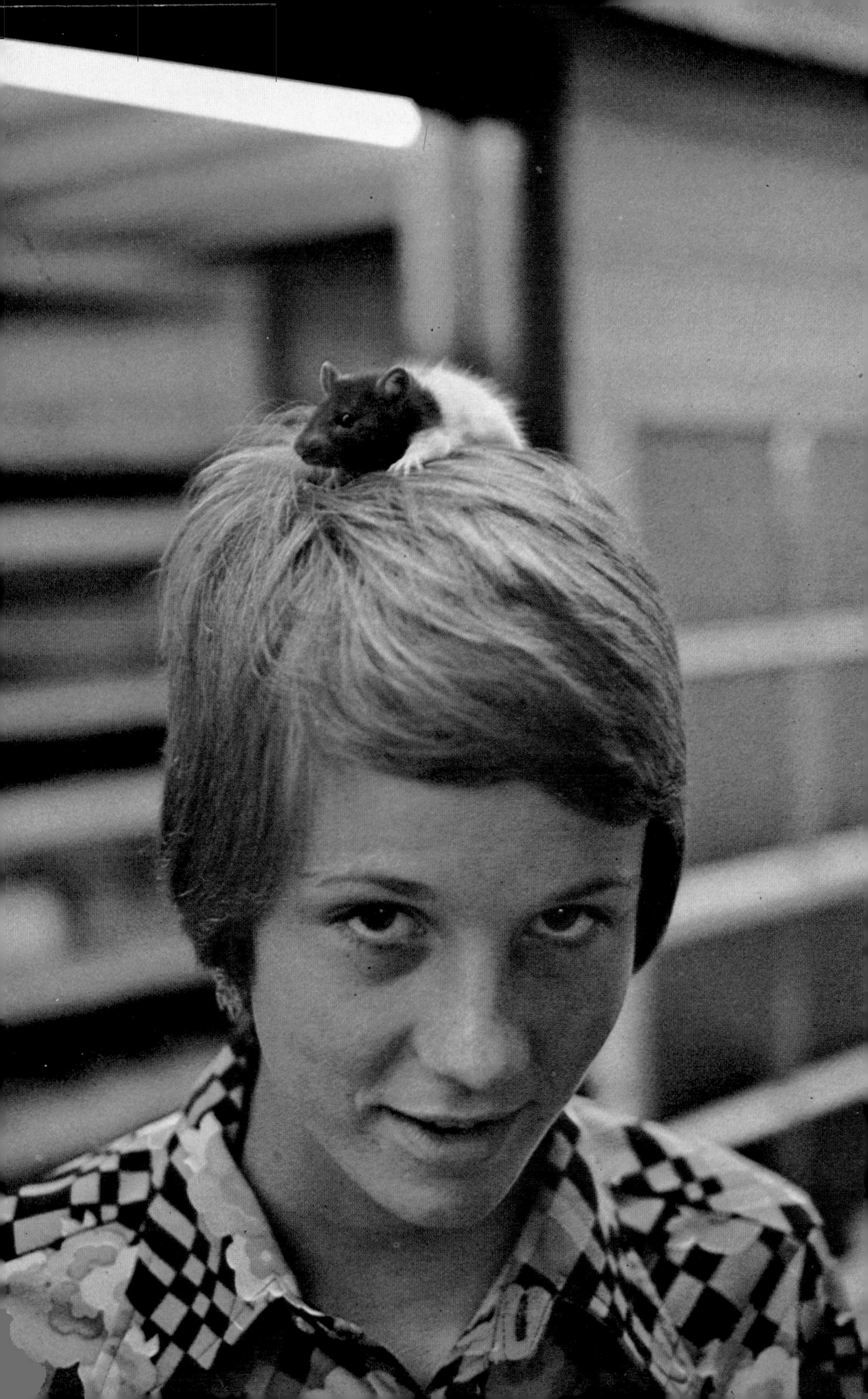

	Mating →	*Birth* →	*Weaning*
Mice		20 days	18 days
Rats		21 days	21 days
Rabbits ...		31 days	42 days

More specifically, the following table provides a relative notion of just where mice and their vital statistics stand in comparison with other small animals and *their* vital statistics.

	Mice	*Rats*	*Rabbits*
Gestation (days)	17-21	19-22	28-31
Estrous cycle (days)	4-5	4-5	rabbits are induced ovulators
Litter size (individuals)	12	12-13	5-12
Age when breeding stops (months)	9	9	24
Number of babies produced by each female in 100 days	21-35	21-28	30

Genetics

The pet owner can control to some extent just what offspring are born. To do this requires some knowledge of genetics—the science of genes and their interactions. Genes, being responsible for the inheritance of characteristics or traits (short hair, pink eyes, agouti-colored coat, etc.), can be *selected for* by the owner. That is, he can pick out parents with the desired characteristics, and then let these mice mate. With a little experience, undesirable characteristics can be *bred out* by not letting the animals with those characteristics mate. In general, one can breed mice in three ways: random mating, the harem system, and the monogamous pair system.

The random mating system is a free-for-all, or colonial system. This method does not require the keeping of any exact records and requires only that one take the best-looking male (that is, the most virile-looking one) and put him together with some of the best-looking females of the same age group. This method allows large numbers of animals to be produced and requires hardly any paper-work at all.

The next system or the harem system, is where one "pasha" is placed together with his bevy of females. The number of females may range from two to twenty. Some experimentation is necessary to use this system because not all species will mate in such a situation. The harem method requires very little space and the resulting number of mice born is very great. One should be aware, however, that in the harem system newborn mice may be smothered by their older littermates or even by the adults.

In the monogamous pair system one encourages nice little "married" pairs who are kept apart from the others. Individual mating records are possible in this system and this, of course, involves paperwork. The detailed records system is only one disadvantage of this method. Much space and attention and equipment is needed for raising even a small colony.

One should be aware of *linkage* when selecting for certain characteristics. Linkage is the tendency for a group of genes to be inherited together continually from one generation to another.

The following section on linkages in house mice should not only give an idea of how your pet mouse could look, but also may explain some "peculiar" mannerisms in your pet. Your pet may be that way because it is its genetic heritage!

One day old mice have no eyes and no ears and no hair. . . but they do have whiskers! Photo by Claudia Watkins, Ferguson, Mo.

An example of linkage would be the albino mouse in which pigmentation is absent in the hair as well as in the iris of the eye. The mouse thus looks white and its eyes are pink, pink being a natural color of the blood in the capillaries and not covered by any pigmentation.

Waltzing mice are abnormal. . . but not sick. . . and they have "different" waltzing patterns. Waltzers come in all colors and make excellent pets. Photo by Mervin F. Roberts.

This book has presented many kinds of mice. Is there an "ideal" mouse? What, in general, does a mouser (that is, a mouse fancier) look for over and above a particular color variety? What is the "ideal" mouse for the breeder? An ideal mouse, according to the British Mouse Club is as follows:

"The mouse must be in length from seven to eight inches from tip of nose to end of tail, with long, clean head, not too fine or pointed at the nose; the eyes should be large, bold and prominent; the ears should be large and tulip-shaped, free from creases, carried erect, with plenty of width between them. The body should be long and slim, a trifle arched over the loins, and racy in appearance, and the tail (free from kinks) should come well out of the back, and be thick at the root or set-on, gradually tapering like a whiplash to a fine end, the length of same being about equal to that of the mouse's body. The coat should be short, perfectly smooth, glossy and sleek to the hand. The mouse should be perfectly tractable and free from any vice, and not be subject to fits or similar ailments.

Sunken eyes, kinked tails, or fits to be penalized . . ."

To keep and maintain close to ideal mice, feed them well, but do not overfeed, if you are breeding for *type,* that is, a particular variety with well-defined characteristics (see the ideal standard above). Obese mice may be sterile. Nursing does, however, as mentioned earlier, should consume all the food they desire.

This is a closeup of a healthy, glistening waltzing mouse. Waltzing mice are not sick. . . they are missing the nerve which makes other mice ill when they try to spin. Photo by Mervin F. Roberts.

SHIPPING

Use stiff cardboard boxes with perforated metal screen portholes on several or all sides. For trips which last more than six hours, light wooden boxes are very appropriate, all of one side being screened. Lightweight metal boxes with ventilating holes on all sides are fine for air shipment of mice.

Wire cages may be inadequate for shipment because darkness is preferred by the nocturnal species, and privacy is very important if the mice are to be kept in good condition. Also, wire offers no protection from extremes of heat and cold.

Before shipping mice, have them in good, well-fed condition. Place foods which have a high water content in the shipping container. Such foods are potato, carrot, apple and lettuce.

Use wood shavings or straw as litter in the shipping container. Arrange for the fastest carrier, and avoid overcrowding if the mice are being shipped in hot weather.

Mice which weigh fifteen to twenty grams should be provided with three square inches per animal in a box five inches high. Not more than twenty-five animals should be in the box.

→

Mice are bred in many colors and with an almost unlimited number of genetic combinations which produce not only different colored animals, but animals with different abilities to ward off disease or grow cancer. These abilities make them desirable laboratory animals. Strains must be kept separate or they will interbreed and ruin their genetic makeup. Photo by Harry Lacey.

For mice which weigh twenty to thirty-five grams, provide four square inches per animal in a box five inches high, and do not pack more than twenty-five mice.

Another way to figure the optimal shipping space is as follows:

Age	*Minimum space*
Mice up to 5 weeks old	3.0 sq. in. per mouse
Mice 5-8 weeks old	4.5 sq. in. per mouse
Mice 8-12 weeks old	6.0 sq. in. per mouse
Mice over 12 weeks old	7.5 sq. in. per mouse

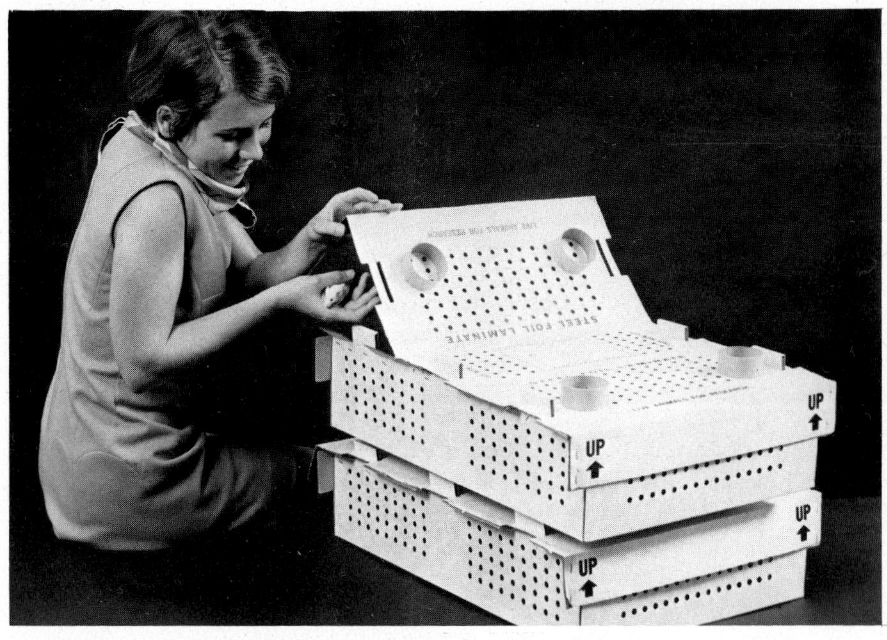

A high-strength shipping container for mice. It has a corrugated center reinforced with steel foil. Photo courtesy of Negus Animal Container Co., Madison, Wisconsin.

When receiving mice which have been shipped, assume that they have been affected by changes in environment (psychological as well as climatic) and diet. Overcrowding, cold drafts, noise, and exposure to infections *en route* all may take a toll. Examine your new animals at once.

Isolate a new mouse for about two weeks before letting it mix with the stock already in your home.

PROFITABLE COMMERCIAL ASPECTS OF RAISING ANIMALS

Mice, along with rats and rabbits, constitute the major percentage of animals used in research; in the U.S.A., for example, mice accounted for about 65% of vertebrates used in laboratories several years ago, rats about 22%, and rabbits about 1.1%. Laboratories use animals for research into the cause and cure of cancer and tumors, for testing the harmlessness (or dangers!) of food color or flavor additives, and for testing drugs intended for human consumption among other things. They are also fed to snakes, birds and other carnivores.

The thalidomide incident—when babies with limb deformities were born to mothers who had taken this drug during pregnancy—inspired a greater testing of new drugs in animals before the release of these drugs for use in human beings. Thalidomide was found to cause much the same teratogenic effect (abnormalities and malformations) in pregnant rabbits as it did in human beings. Rabbits, therefore, are being used quite extensively for this purpose, although rats are now beginning to contribute, too.

Large numbers of rats, rabbits, and mice, as well as other small animals, are used in the bioassay of literally thousands of prophylactic and therapeutic substances before the U.S. government will permit those substances to be released for sale to the public. Universities, research institutes and foundations, cancer research units, hospitals, U.S. Public Health laboratories, and the pharmaceutical industries are the organizations which use these animals for the above purposes.

Mice, rabbits and rats are also used for the laboratory diagnosis of diseases and of pregnancy in human beings.

Mice make great pets that's why 2,000,000 are sold through pet-shops every year!